I0821707

THE SPURS OF JAMES J. Wheat

PIONEER COLLECTOR

BY BRUCE BARTLETT
PHOTOGRAPHY BY BRANDI PRICE
EDITORIAL SUPPORT BY MARSHA PFLUGER

PUBLISHED BY THE NATIONAL RANCHING HERITAGE CENTER
LUBBOCK, TEXAS

ISBN 9761834-6-3

FIRST EDITION

Published by the NATIONAL RANCHING HERITAGE CENTER, Lubbock, Texas, USA

With underwriting from the DIAMOND M FOUNDATION, San Angelo, Texas

Graphic Design, Art Direction and Photography by Brandi Price and Mark Hartsfield, Hartsfield Design, Lubbock, Texas

Editorial Support and Project Direction by Marsha Pfluger, National Ranching Heritage Center

Printing by Craftsman Printers, Lubbock, Texas

DEDICATION

This book is dedicated to my mom and dad, Ruth and Ernest Bartlett, who instilled in me from an early age a love of the American West that still burns bright today.

To my wife, Julianne, without whose support, love, steady hand and hard work this book would not have been possible.

To my son, Winn, whose embrace of all things Western inspires confidence in the next generation.

To the preeminent collectors, historians, authorities and close friends Kurt House, Ellis Ramsey, Jim Statler, Randy Butters and Joe Flores who have added so much to my appreciation and understanding of the field of spur collecting.

And lastly to the dedicated staff at the National Ranching Heritage Center for allowing me the opportunity to participate with them in this venture.

K&C

TABLE OF CONTENTS

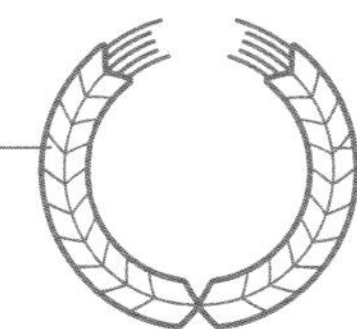

JAMES J. *Wheat*

LEGEND AND REALITY

James J. Wheat was from a time when oil made legends of average men. As the prominent TV journalist and author Charles Kuralt wrote in his "On the Road" book, "Loving County is the emptiest county in the United States. It covers 669 square miles and has a population of 91 persons. That's about one lonely soul for every seven square miles." It is in that place in the town of Mentone where James Jackson Wheat resided and grew rich on the local oil, which was about the only thing that grew there.

J.J., as he was known, was a character. He was born in 1916 and left this earth in 1989. His father, who had the same name as J.J., brought in his first well in 1924. He had wildcatted several more productive wells, leaving his heirs very wealthy. J.J., being one of those wealthy heirs, drove a series of Rolls Royce and Mercedes-Benz automobiles, mainly to move his herd of 300 cattle across his rugged property in Loving County. Although it rarely rained and there was no other good source of water, the oilman had it hauled in, because as he explained, he needed to keep a few cattle around for a tax deduction.

Over a span of 10 years, J.J. acquired scores of Texas Style spurs, including the largest-known group of rare Swede Strong spurs known to exist. He mainly relied upon a couple of knowledgeable dealers in Texas and Oklahoma to locate bits and spurs for him. And when they did, J.J. paid them for their finds with cash he pulled from his stuffed pockets.

Carl Jennings of Tulia, Texas, was one of Wheat's most trusted buyers. Jennings said that toward the end of the collector's life, Wheat asked him what he should do with his bits and spurs. Since Wheat's family did not share an enthusiasm for spurs, Jennings recommended he donate the collection to the National Ranching Heritage Center. In November 1985, Wheat gave the ranching museum 920 spur pairs and singles and more than 100 bits. This book shows many of the best pieces from that collection.

As assessed by spur expert and dealer Bruce Bartlett of San Antonio and with assistance to the staff from Joe Flores of Stratford, Texas, the Wheat Collection was found to hold dozens of prized spurs from Texas Style makers. The Swede Strong group is the largest of the spur maker's collected work that either man has seen. The collection reflects an evolution from well-worn cowboy spurs to rare, collectible Western Americana. Through it, people can learn about spur styles and patterns and the artisans who created them.

Knowledgeable spur collectors advised the NRHC to document the best of the Wheat Collection in a photo book. The project was undertaken, and through the Diamond M Foundation, funding was acquired to make this the third book from the publications division of the National Ranching Heritage Center.

The Wheat Collection represents just one of the ways in which staff and supporters preserve the past and educate others about many aspects of the *real* West. The NRHC was established to maintain the colorful history of ranching and pioneer life in the American West. Since its public opening in July 1976, that mission has been carried out.

During a visit to the National Ranching Heritage Center museum and historical park, people can see:

- 30 pieces of life-size, outdoor, bronze sculpture, which enhance the grounds of the museum.
- Exhibits of art, photography and artifacts.
- More than four dozen rare ranch structures, including a mail camp, bunkhouse, one-room schoolhouse, dugouts, ranch commissary, cookhouse, windmills, ranch homes and offices, a depot, train and shipping pens. All were relocated from real working ranches of the past or present and restored to reflect life from 1780 through 1950. Each structure is authentically furnished or outfitted to represent its primary years of use.
- The museum gift shop, designed in the style of an old general store.

Special events and educational programs are presented throughout the year. Memberships are offered in the Ranching Heritage Association, the primary support organization for the National Ranching Heritage Center. The museum and historical park are open Monday – Saturday, 10 a.m. to 5 p.m. and Sunday, 1 to 5 p.m.; closed on major holidays. Visit the NRHC online and check out the podcasts and virtual tours at www.NRHC.com.

By Jim Pfluger, Executive Director
RANCHING HERITAGE ASSOCIATION and
NATIONAL RANCHING HERITAGE CENTER
Lubbock, Texas

INTRODUCTION

A PERSONAL LOOK AT THE WHEAT COLLECTION
BRUCE BARTLETT, JUNE 19, 2008

I have had the honor of handling, admiring and appraising each bit and pair of spurs belonging to West Texas pioneer collector James J. Wheat. He, like fellow collectors of the era Tom Reagan of Beeville, Texas, and O.R. Huff of Fort Worth, found collecting spurs to be a daunting task. Prior to the founding by Bill Bonser and others of the National Bit, Spur & Saddle Collectors Association in the mid-1980s, there was no organized group committed to bringing together individuals who had an interest in the field.

Too, before field-specific books and shows developed, the only way collectors could learn about spurs and bits was through talking to the old-timers who remembered the actual makers and their products and gleaning what they could from the spurs they were able to collect. The actual collecting was done by purchasing or being given spurs by their original owners.

Most consider the Golden Era of the Texas spur-making tradition to be roughly 1910 to 1940 with the high-water mark being the late 1920s. Prior to 1900, the Texas Style spur had not fully evolved as a distinct regional style with its own characteristics. There have been many fine books written on the Texas Style spur, which 98 percent of the Wheat Collection is composed of. In my opinion, the best book ever written on the subject is the "Bit and Spur Makers in the Texas Tradition" by Ned and Jody Martin and Kurt House. The book covers fully the evolution of the Texas Style spur and its characteristics. It discusses in the greatest detail yet published the most extensive list of the field's Early Masters.

The main purpose of the Wheat book is to add to what has already been documented so impressively by the many fine books on spur collecting that have been written since the mid-1980s. This book focuses that information and more on the spurs and bits collected by one man.

Spur collectors then, as today, are "accumulators" at heart — the main difference between then and now being the access to information. With information came sophistication. Wheat must have known very little about the history of the makers he collected or the patterns they produced, both of which have been carefully researched today. In short his collection was limited to the information available to him at the time.

Today's advanced collectors normally focus their interest in spurs of a specific region. In the United States, there are three main regional styles that have developed since 1890 — the Texas Style, California Style and Plains Style. If a collector is from Texas, he most likely is interested in Texas Style spurs. If a collector is from California or the Far West, his attention is typically directed at the California Style.

Each regional style has its own unique set of design characteristics. The collector may choose to acquire a few examples of each style by all the Early

Masters or just focus on one or two of the best-known regional makers. Normally a plan develops over time. As the collector further hones his area of interest, the spurs and bits are usually upgraded, with lesser or non-conforming examples being sold or traded off. Over time the collection, considering budget, will comprise only the best examples obtainable.

Collections assembled before 1985 share many common characteristics with those today.

COLLECTORS 2008

- Collect nationally with an extensive show, auction and dealer and collector network
- Small numbers of spurs of the highest quality available
- Restored where applicable
- Must be constantly on the lookout for all types of forgery
- Numerous great books for reference
- Many unmarked local and regional makers identified
- A very expensive pair: $30,000

EARLY COLLECTORS PRE-1985

- Collected what was available regionally; no national market
- Collected large numbers of spurs of all qualities with only a small percentage being in the "great" category
- Collected mostly unrestored examples
- Rarely encountered fakes; if present, they were crude attempts
- Spurs were obtained at flea markets, antique stores, local auctions, from individuals and a limited number of specialized dealers. Little information was available, no field-specific books, only marked examples, easily identifiable
- A very expensive pair: $1,000

Over the years, I have been invited to view and many times purchase collections and individual pieces acquired prior to 1985. These collections were seldom assembled with any special considerations given to condition or patterns. The Wheat Collection fits this category. Wheat, like his fellow collectors of the period, bought without apparent regard for most of the guideposts used by today's collectors. In Wheat's time, a very expensive pair of spurs would have been $1,000, making collecting a much lower stakes proposition than today, when an outstanding pair can exceed $30,000. Spur collecting was a much different endeavor in those days. As larger values and more information have entered the field, so have forgery and restoration. The Wheat Collection contains minimal early examples of forgery that would fool few collectors today. For the most part it is a pristine, unrestored assemblage of what was available at the time.

Wheat and his collecting contemporaries would have found additions to their collections through a handful of specialized dealers, antique stores, flea markets, local auctions and word of mouth. He also obtained items from the families of the purchaser or in many instances from the original purchaser or maker himself. Wheat was by all accounts a colorful man of means, and we can assume by the descriptions of those who knew him that he could afford whatever he wanted.

In those days, a collector had to take what came along, explaining why of the many hundreds of pairs in the Wheat Collection, fewer than 100 examples would be considered truly "great" by today's top tier of collectors. Wheat would have probably chosen to possess the work of makers from his own region like the collectors of today, but the lack of a regional and national marketplace for collector spurs would have left him little choice. He collected what he could find locally — old and new.

Wheat was very lucky, however, because in the 19-teens and '20s, Harold "Swede" Strong operated out of Pecos, Texas, very near Wheat's ranch. As a result, the collection has the best assemblage of Strong's work known to exist. J.O. Bass was also nearby, in Tulia, Texas, to the north, so Wheat was able to accumulate a large number of Bass spurs, as well. He also was interested in L.G. Grubb and Carl Hall who were making spurs at the time.

From the many fine, unrestored examples of the Early and Middle Period Masters found in Wheat's collection, much information can be gleaned using the resources we have today. This book offers a rare opportunity for collectors and enthusiasts who have seen few unrestored spurs to witness first-hand the average condition of so many early Texas spurs.

The Wheat Collection has been divided into broad categories represented by chapters in this book.

MEXICAN CHIHUAHUA

BEGINNING OF THE TEXAS STYLE

BEGINNING OF THE TEXAS STYLE

Mexican Chihuahua Spurs
Military Spurs
August Buermann

The American non-military spur-making industry began to emerge around the 1870s. Prior to this time, cowboys used mostly Mexican or Military Style spurs. As the spur industry developed, two distinct styles appeared. The California Style and Texas Style. Both were influenced by the desires and needs of the customer: Western cowboys.

The California Style and Texas Style were strongly influenced by the prevailing spur making trends in Mexican states, south of their respective borders. The spur makers from northeast Mexico (below Texas) in the 1860s favored larger, one-piece spurs with wide heel bands and shanks.

The makers from northwest Mexico (near California) in the 1860s had as their primary influence the European and Spanish Colonial Styles. These regional preferences were carried across the border and became the foundations on which the California and Texas Styles grew. Aside from these two distinct traditions, there was a third that mixed the design characteristics of the California and Texas Styles. The result became knows as the Northern Plains Style spur.

CIRCA 1880 – 1930 MEXICAN CHIHUAHUA SPURS.
(Above) Most pre-1900 cowboys in Texas wore this style of spur originating in the Mexican states of Chihuahua and Sonora. The Mexican spur-making tradition had a large impact on the Texas Style as it began to take shape in the first years of the 1900s.

A SAMPLING OF 19TH CENTURY AND EARLY 20TH CENTURY MILITARY SPURS

Many individuals who served as horse soldiers during the Civil War and in other campaigns returned home with their spurs. These military designs had a lasting impact on Western cowboy spur styling.

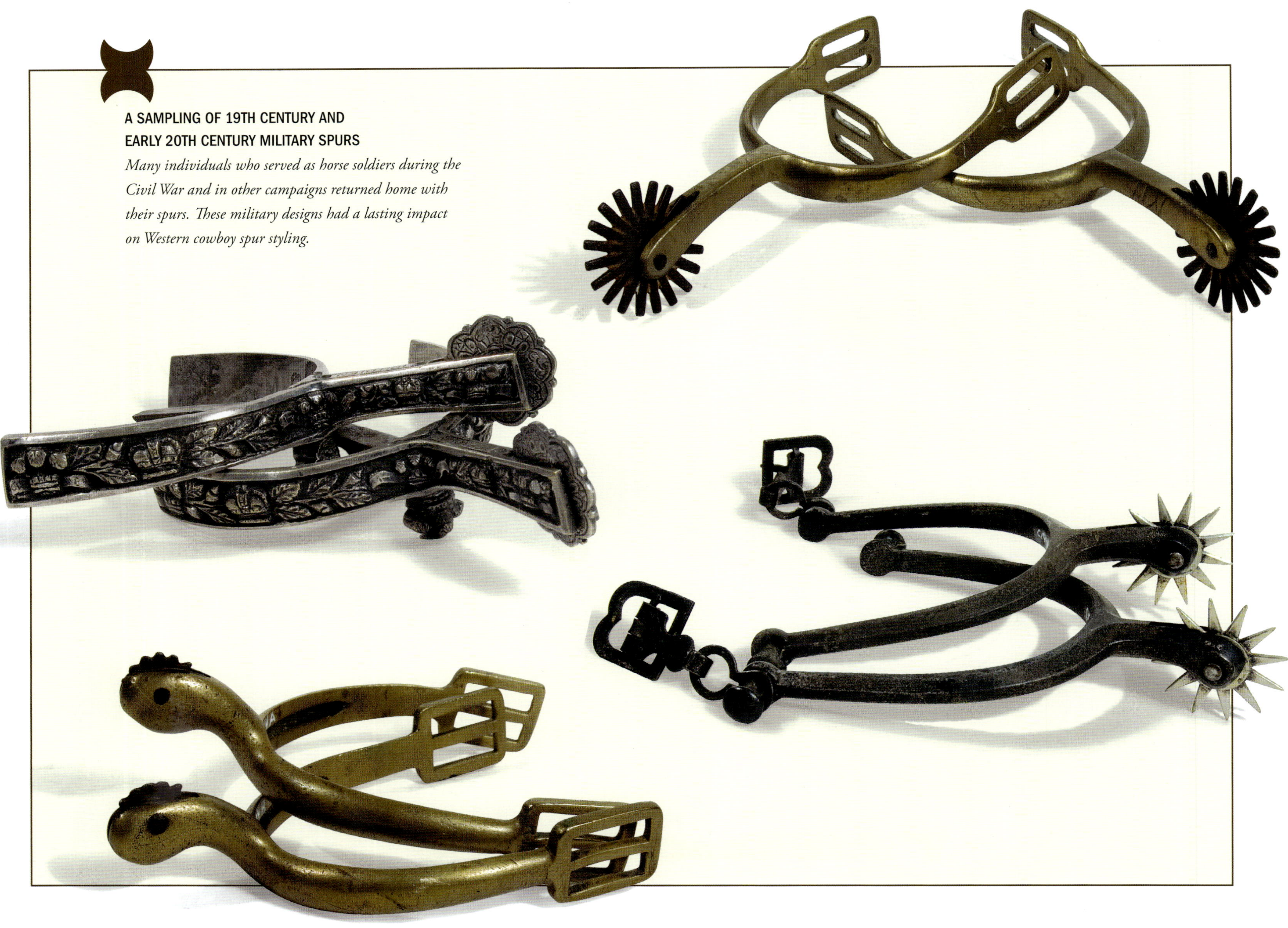

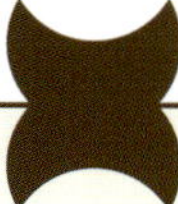

BUERMANN

No other maker did more to jumpstart the American cowboy spur-making tradition than August Buermann, who began fabricating spurs in 1866. These two early designs enjoyed a long production run.

AUGUST BUERMANN (1842-1928)

A native of Hanover, Germany, August Buermann came to America in 1863. He apprenticed with a saddlery hardware manufacturer, Alexander Barclay & Co., in Newark, New Jersey. He bought the company in 1866 and in 10 years employed 35 skilled workmen. Though headquartered in New Jersey, Buermann manufactured California, Mexico and Texas spurs, employing many of the finest craftsmen from those regions. Orders came from horsemen throughout the world because of Buermann's patented features on bits and spurs. In 1902, the company was moved to a larger, four-story brick building, which manufactured 443 different spur designs at the peak of Buermann's popularity. An astute businessman, he brought his sons into the Buermann Manufacturing Co. business and expanded the lines to include marine and automotive hardware prior to World War I. His inventive genius resulted in many registered patents. The company sold in 1926 to a competitor, North & Judd, whose owners kept the Buermann designs in production for another decade.

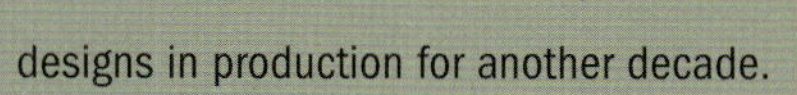

AUGUST BUERMANN

BUERMANN'S "OK" SPURS

This venerable pattern (left and below left) was produced for 40-plus years and perhaps more than any other spur has come to symbolize the humble origins of the American cowboy spur.

BUERMANN

August Buermann's bull head pattern with arrow shanks, circa 1920s. The bull head was made by a number of the period's makers. One of the Buermann company's great strengths was to identify what its competition was selling well then make a comparable product to monopolize on its success.

J.O. BASS

THE GREAT COLLECTIONS

THE

GREAT COLLECTIONS

Harold "Swede" Strong
James Oscar Bass
Robert Lincoln Causey
George A. Bischoff
John Robert McChesney
Oscar Crockett
Pascal M. Kelly

James Wheat was able to collect a significant number of examples of seven of the Texas Styles' most important Early Masters. It is not known whether Wheat attempted to collect all these makers in the numbers he did although it is known he specifically sought out Harold "Swede" Strong. By today's standards, a great collection of each of the seven makers in this chapter represents a different quantity and quality of spurs.

Broadly, any collection today that contains a single example of Swede Strong, J.O. Bass, Robert L. Causey or G.A. Bischoff is an exception rather than the rule. To have all four makers represented is virtually unprecedented.

The collections of these four makers is further remarkable because of the sheer number of pieces of their work represented. For example, the nine Harold Strong spurs and bits is the largest accumulation of his work known.

Although more obtainable, Wheat's collection of Early Masters J.R. McChesney, P.M. Kelly and Oscar Crockett is exceptional. Although these makers' products are much more common today, Wheat's collection contains enough examples of each in rare patterns and excellent condition to make these collections significant.

HAROLD "SWEDE" STRONG

Two fine pairs of flying S-marked Harold Strong spurs. The pair (top) with Marfa shanks was made stylish by cowboys in the Big Bend Country, south of Pecos.

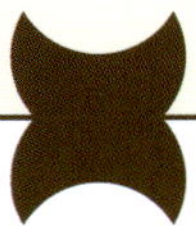

HAROLD "SWEDE" STRONG

Outstanding, matching Strong-marked gal-leg spurs and bit. Of the seven pairs of spurs in the Wheat Collection, all have swinging buttons.

HAROLD "SWEDE" STRONG (1887-1970)

Born in Sweden, Harold Strong ran away after his mother died. He was strong, tall, intelligent and big-hearted. A teenager at the time, he worked his way from New York to Chicago and was hired as a stone mason by a company that built the Texas State Capitol building. Swede became a citizen so he could enlist for duty in WWI. There, he received training as a blacksmith. After the war, he came to the Oliver Loving ranch near Rotan, Texas, asking for a job as a cowboy. He became like a big brother and protector to Loving's great-granddaughter, Mrs. Mabel Brown. He enjoyed photography and leather work, later operating the Pecos Saddle Shop. He created two saddles that won grand prize at the Pecos Rodeo. Some say he made 49 pairs of spurs in his life. Others say he made 49 pairs of gal-leg spurs. Either way, one of his biggest patrons was J.J. Wheat, a Loving County oilman and spur collector. In Swede's final years, Mrs. Brown took care of him and had a military veteran headstone placed on his grave in the Brown family's plot in Baird, Texas.

HAROLD "SWEDE" STRONG

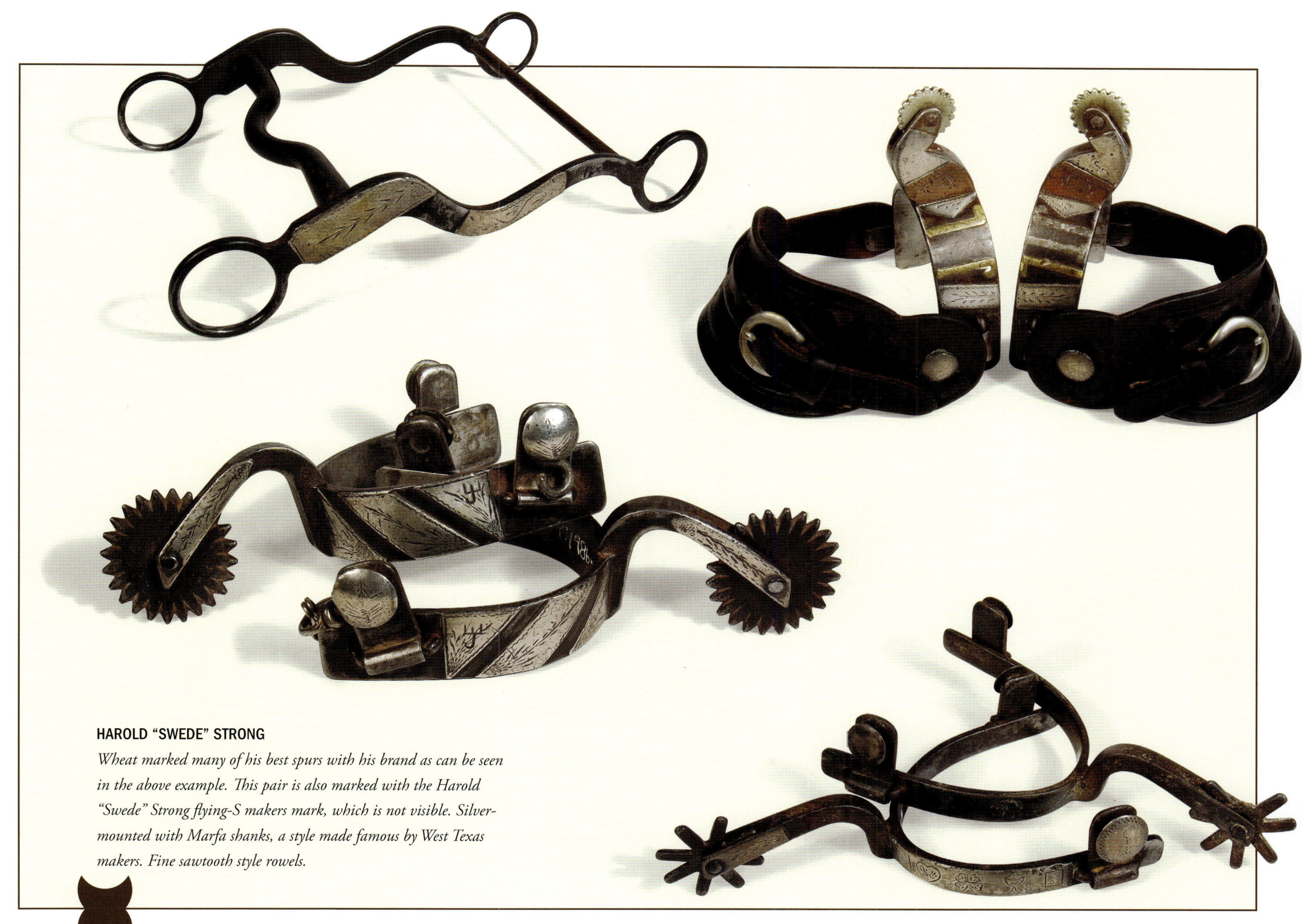

HAROLD "SWEDE" STRONG

Wheat marked many of his best spurs with his brand as can be seen in the above example. This pair is also marked with the Harold "Swede" Strong flying-S makers mark, which is not visible. Silver-mounted with Marfa shanks, a style made famous by West Texas makers. Fine sawtooth style rowels.

J.O. BASS

B

JAMES OSCAR "J.O." BASS (1879–1950)

J.O. Bass was born in Atlanta, Georgia, and moved to Texas at age 11. He opened the first blacksmith shop in Quitaque, Texas, at age 18. His one-piece spurs and bits became so popular with cowboys that he gave up all other work to fill their orders. In 1905, with his bride, Corrie, Bass moved to Tulia, Texas, where he made bits and spurs to the buyer's order. Soon, the Texas Rangers began using his spurs, as well as rodeo stars and such cowboy personalities as Tom Mix. On many of his spurs, he used a half-silver, half-copper heart-shaped button, which became a Bass trademark. His first catalog was issued around 1909. Bass was known as a kind, well-liked man who was generous with his expertise, freely teaching others the art of spur making. In 1924, he made his last pair of spurs and quit the business to farm. He moved to Plainview in 1938 and died of cancer in 1950. The desire for his bits and spurs has created some of the highest prices paid by collectors.

J.O. BASS

(Right) A heart pattern with classic six-point locking spoke rowels. The best pair of Bass spurs in the collection, these feature unusual spade swinging buttons and a one-of-a-kind heart motif created especially for the purchaser.

J.O. BASS

West Texas master J.O. Bass made all these spurs, one of the largest assemblages of his spurs known. Bass's penchant for card suits is clearly evident in the pairs he produced across the entire 20-plus years of his spur-making career. His spurs were ordered by cowmen who bought them to use. Like most Basses that have been found, these all show various tell-tale signs of excessive use; some have replaced rowels, some are missing silver mountings, and most have wear in the contact points of the swinging buttons. It is extremely difficult to collect Bass spurs in excellent condition, as few pairs escaped use. The spur maker numbered each pair sequentially beginning with his first pair. He made multiple examples of his two most popular designs, the 217 and 400 patterns, which also appear in his only known catalog. J.O. Bass kept a ledger that recorded each bit and spur he made including its description, price and purchaser. This ledger, his unique makers mark and quintessential Texas Styling have made J.O. Bass the most collectible of Texas' Early Masters. In his career, he made fewer than 2,300 pairs ranging in price from $4 to a princely $18 for his most elaborate designs. Today, an extraordinary example could sell for $30,000 or more.

R.L. CAUSEY

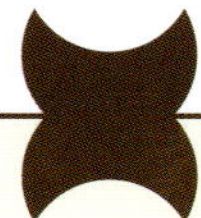

CAUSEY

Fewer than 10 marked pairs of Robert L. Causey spurs are known to collectors, making this pair an extreme rarity.

ROBERT LINCOLN "R.L." CAUSEY (1868-1937)

R.L. Causey ran away from his farm home of 10 brothers and sisters in Kansas City, Missouri, when he was just 12 years old. With $5 that he had borrowed from a neighbor, the young man took the train West as far as it would go. In Indian Territory, he found a small community with a blacksmith shop. His father had taught him some blacksmithing, so he worked for room and board and $1 a week, repaying the $5 loan. Causey remained there for four years then traveled to Lovington, New Mexico, to see his brother, George, a buffalo hunter-turned-rancher. R.L. helped George build a ranch house, corrals and other structures then moved to Odessa, Texas, in the late 1880s, becoming a well-known spur maker. Five years later, R.L. moved to Eddy, New Mexico, and married in 1903. Always filled with a desire to roam, he moved around the country, taking his family along with him. His final home was in Safford, Arizona, where he continued to create beautiful bits and spurs until he died from a kidney infection. Causey is credited with developing the gal-leg and horse-head designs.

CAUSEY

Robert L. Causey's shanks, heel bands and buttons were quite thin, as can be seen in these views.

CAUSEY

Although unmarked, this pair was certainly made by Robert L. Causey, who incorporated design features from the Plains and California Styles into his spurs. These have Cheyenne Split heel bands with inlaid silver diamonds on classic Texas gal-leg style ironwork.

CAUSEY

Notice the similarity in the thickness and gal-leg design of the shank and boot of these three pairs of spurs.

CAUSEY

Robert L. Causey is one of spur collecting's most difficult makers to acquire. Most of his work left his shop unmarked making identification difficult. The marked pair in the middle allows the trained eye to attribute the two unmarked pairs to Causey.

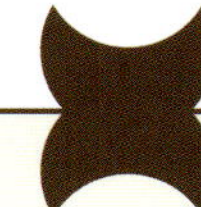

BISCHOFF

Circa 19-teens G.A. Bischoff Style A spurs. Although the least desirable of Bischoff's initial catalog designs, this pair is still extremely valuable. Most collectors of early Texas spurs would consider themselves lucky to have a single Bischoff-made piece in their collection. Note the copper peanut missing from the heel band of left spur. G.A. Bischoff spurs and bits are the most valuable on a per pair basis of all of the early Texas Masters.

GEORGE A. BISCHOFF (1862-1944)

One of 10 children, Bischoff was born in Emmettsburg, Maryland, of German emigrant parents. He apprenticed in a carriage factory and spent the rest of his life working in his trade from coast to coast in 16 states. He married in 1890. In Gainesville, Texas, Bischoff worked for a time with J.R. McChesney, but in 1911 published his own catalog – everything was of one-piece construction, and he was the only spur maker to list dimensions of his products and weight. At this same time, he ran George A. Bischoff & Co., High Grade Bits and Spurs. An innovative spur maker, George was also active in civic affairs, serving three terms as a city councilman. He owned the first Maxwell automobile dealership in town, locating it on the ground floor of his bit and spur business, which was upstairs. When the last livery stable closed, Bischoff sold his business in 1915 to C.P. Shipley and moved to Kansas City. During the 1920s, he worked with machinery in the oil fields and owned the Western Wheel and Body Works with his son, William. Bischoff was a highly skilled tool and die maker for Douglas Aircraft Corp. in Tulsa when WWII began. He was the company's oldest employee at age 82. Bischoff is thought by some to be Texas' premier spur maker. His company only operated for five years, so his bits and spurs are rare and valuable, marked inside the heel band of both. He continued to work at Douglas Aircraft until retiring due to poor health four months before his death.

McCHESNEY

McCHESNEY

Four pairs of ladies' McChesney spurs, circa 1920s. (Starting from the top) No. 22 pattern, No. 23 pattern, No. 23 pattern and No. 23 pattern. Most vintage cowgirl spurs are in excellent condition like these. It could be assumed that ladies found themselves horseback much less frequently than did their male counterparts.

JOHN ROBERT "J.R." McCHESNEY (1868-1928)

Born in Plymouth, Indiana, J.R. attended Notre Dame for two years before moving West, ultimately into Indian Territory. He bought a bankrupt Tulsa blacksmith shop and freighted it to Broken Arrow, Oklahoma. He married in 1884 and made his first pair of spurs in 1887 after "being encouraged to do so" by the gun of a cowboy who really wanted a pair of spurs. With the drought of 1889, McChesney moved the blacksmith business to Gainesville, Texas, where one of his employees was George Bischoff. The business grew rapidly. All seven of J.R.'s sons learned bit and spur making. The family moved to Pauls Valley, Oklahoma, in 1909, where McChesney Spur and Bit Co. became the largest such manufacturer west of the Mississippi River. His catalog featured 120 patterns and the business employed 50 people, including P.M. Kelly and Tom Johnson as finishers. McChesney did not mark his bits and spurs except for placing an arm and hammer on products supplied to C.P. Shipley Saddlery for two years. Mac stayed with hand-forged methods when others moved to machine production. His gal-leg, gooseneck and peacock designs ranked him among the greatest innovators of bit and spur designers. The changing business climate may have caused a fatal heart attack in 1928. J.R.'s sons left spur making and went to work in the oil fields. His wife sold the business to Justin Nocona Boot Co., which marked his designs "McChesney." Twenty years later after Nocona had disposed of the equipment, Adolph Bayers bought the McChesney tools and dies from a junk man.

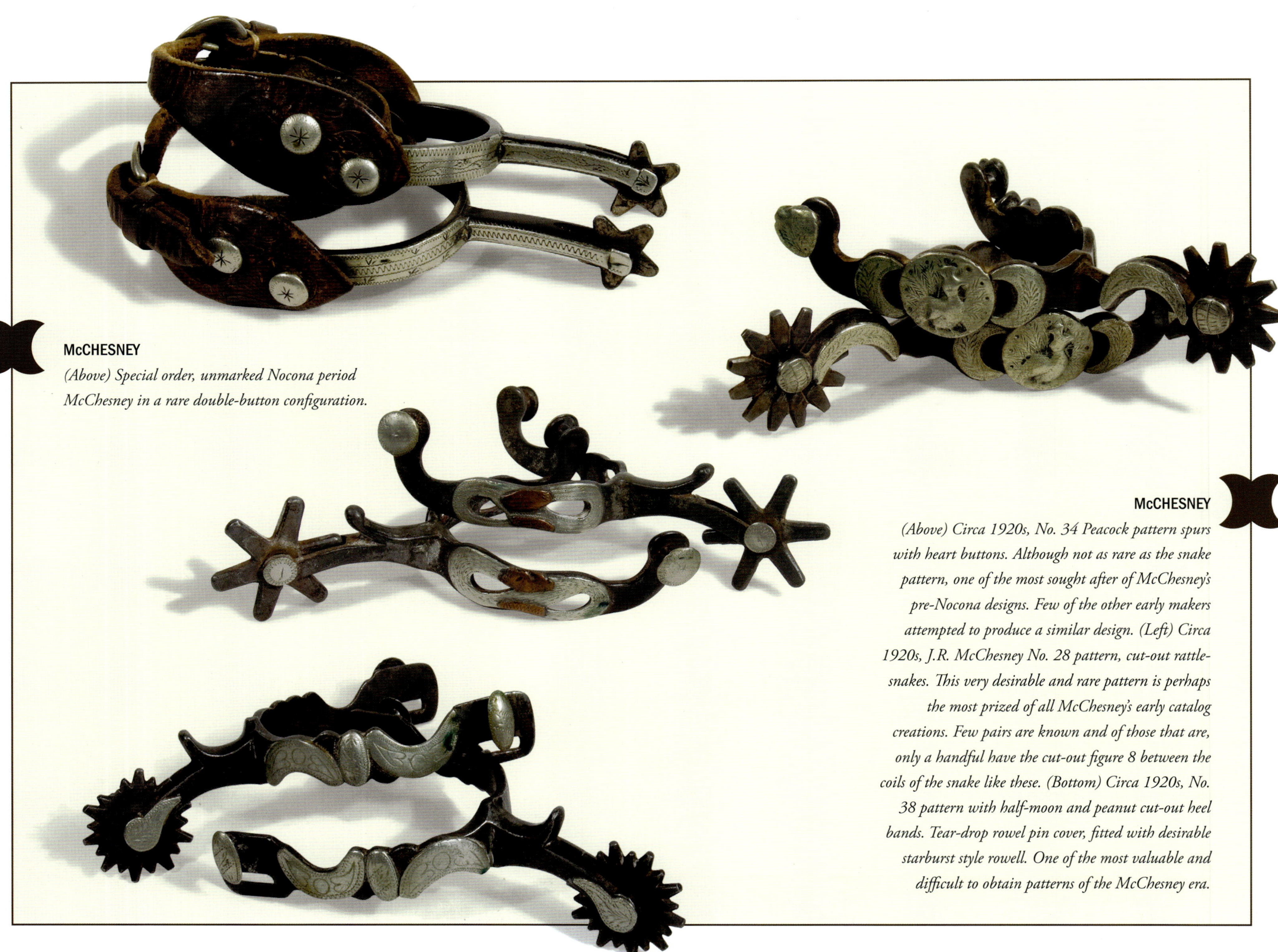

McCHESNEY

(Above) Special order, unmarked Nocona period McChesney in a rare double-button configuration.

McCHESNEY

(Above) Circa 1920s, No. 34 Peacock pattern spurs with heart buttons. Although not as rare as the snake pattern, one of the most sought after of McChesney's pre-Nocona designs. Few of the other early makers attempted to produce a similar design. (Left) Circa 1920s, J.R. McChesney No. 28 pattern, cut-out rattle-snakes. This very desirable and rare pattern is perhaps the most prized of all McChesney's early catalog creations. Few pairs are known and of those that are, only a handful have the cut-out figure 8 between the coils of the snake like these. (Bottom) Circa 1920s, No. 38 pattern with half-moon and peanut cut-out heel bands. Tear-drop rowel pin cover, fitted with desirable starburst style rowell. One of the most valuable and difficult to obtain patterns of the McChesney era.

McCHESNEY

NOCONA McCHESNEY

(Left) No. 70, 3-heart pattern Nocona McChesney nickel-plated spurs, 1930s. (Right) No. 19 pattern Nocona period McChesney ladies spurs. Nickel-plated and in mint condition. Circa 1930s. (Below) Rare Nocona period McChesney steer-head motif spurs. The styling of this spur's ironwork was strongly influenced by the Mexican Chihuahua spur, popular in Texas during the period.

McCHESNEY

Two very unusual, early McChesney bits. The figure 8 pattern coiled snake (right) is among the most prized of all of McChesney's catalog bit patterns.

McCHESNEY GAL-LEG IMPACT ON DESIGN

Among the most popular Texas Style spur motifs ever created was the venerable gal-leg or ladies-leg. It was produced in various forms by virtually every early maker. Widely credited with being the gal-leg's original creator, J.R. McChesney, at the very least made it widely popular. (From left, top) Kelly Bros., blacksmith-made, McChesney. (From left, bottom) blacksmith-made, August Buermann, Oscar Crockett, McChesney.

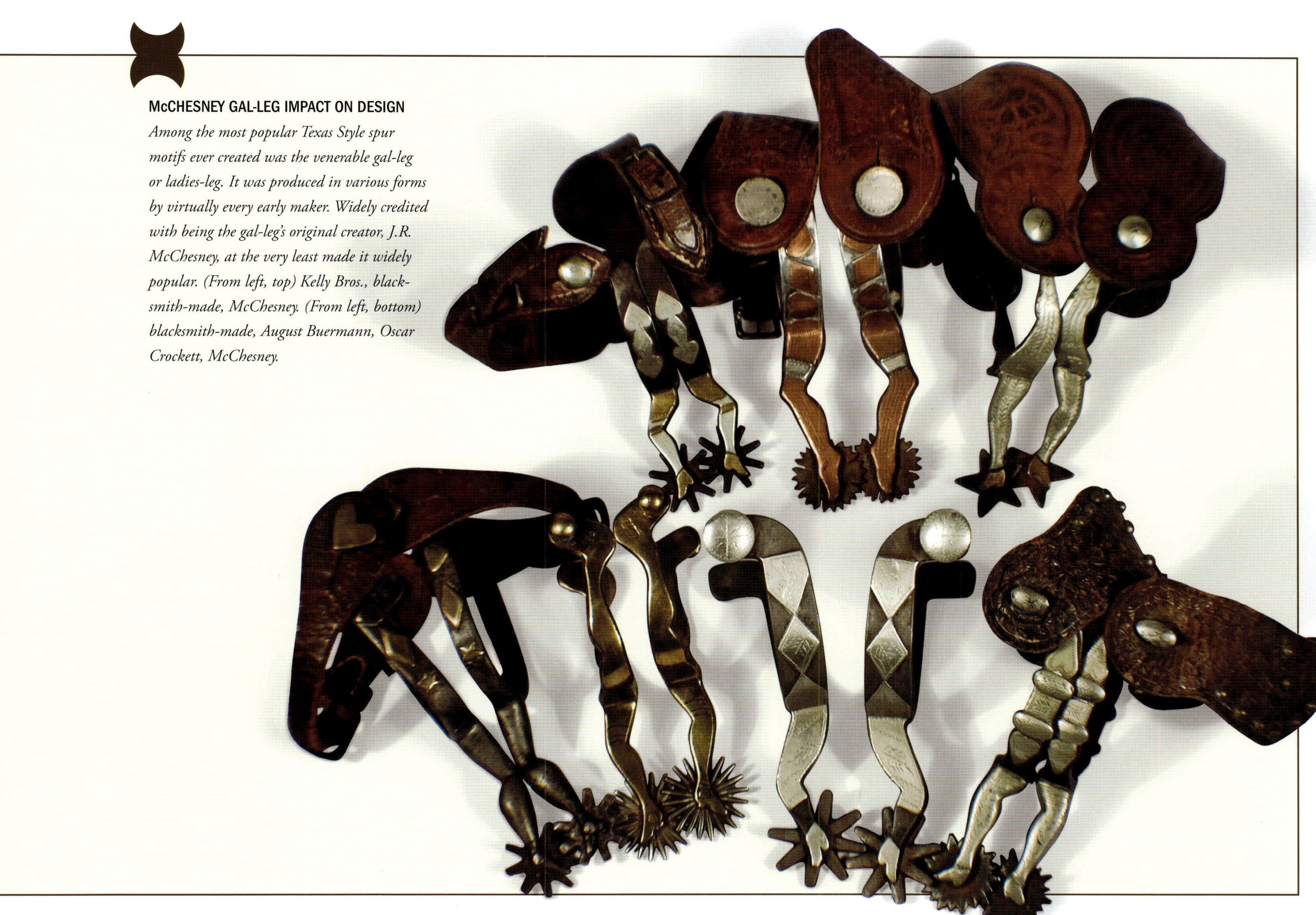

CROCKETT

(Below) Circa 1920s, Oscar Crockett Marfa-shank pattern spurs with peanut mountings. The pins holding the rowels were replaced during their period of use causing the rowel pin mountings to be removed. Cowboys of the era seldom replaced mountings when they were lost or removed.

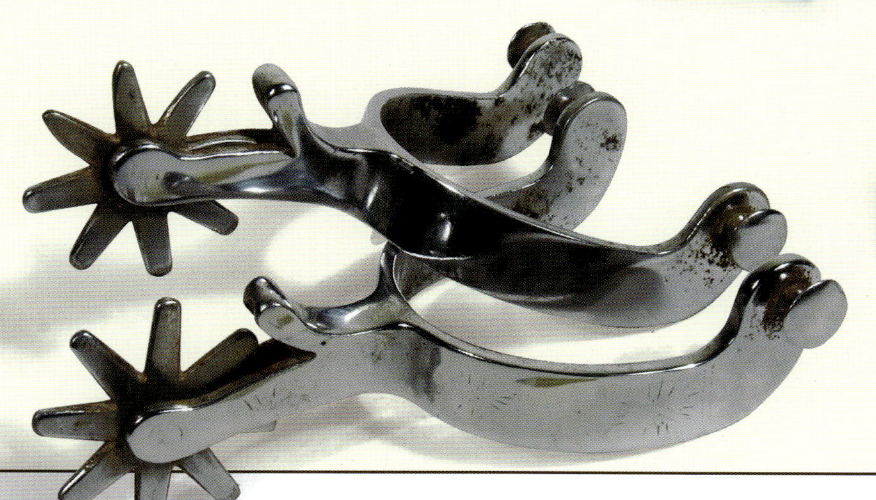

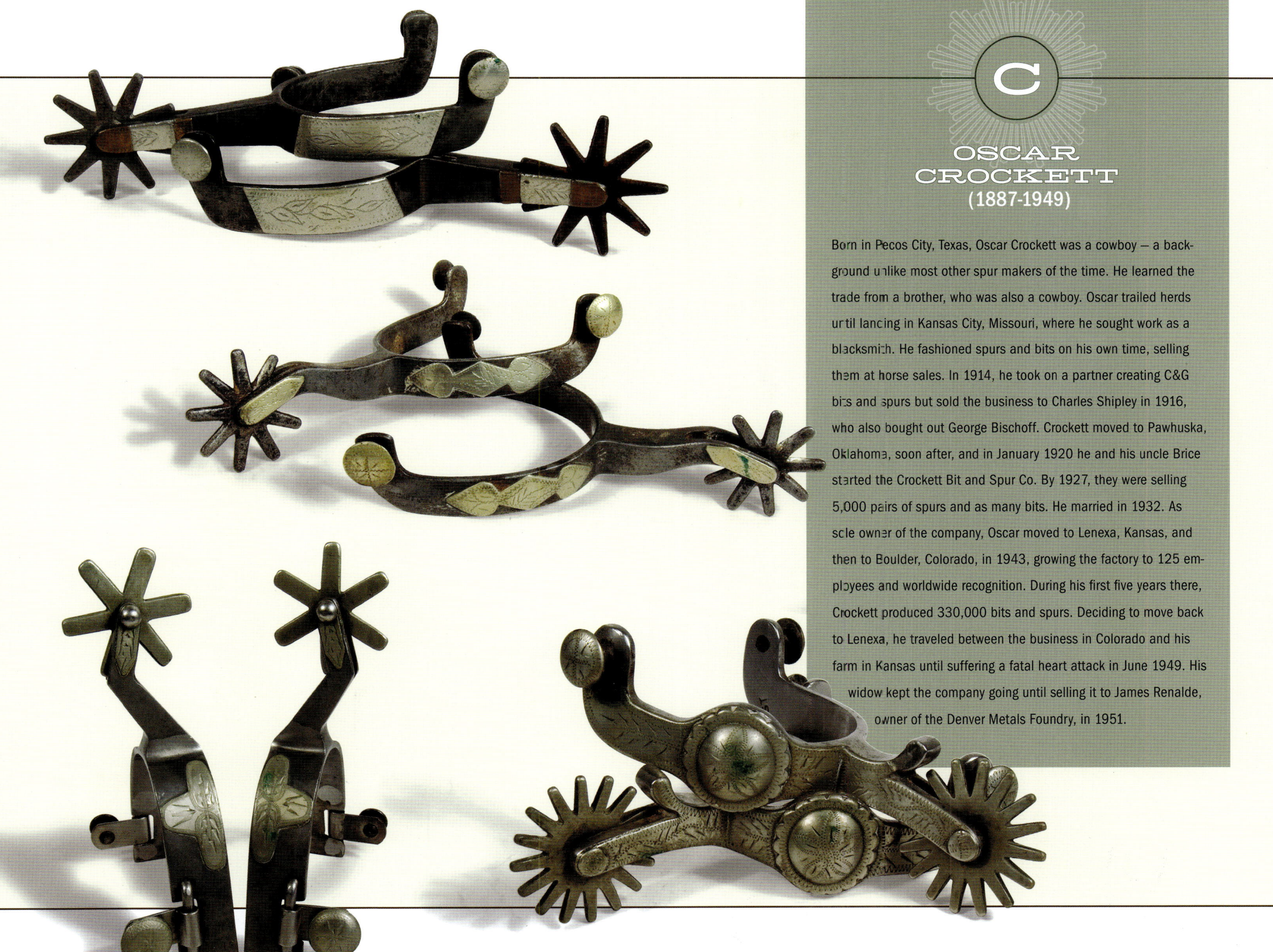

C

OSCAR CROCKETT (1887-1949)

Born in Pecos City, Texas, Oscar Crockett was a cowboy — a background unlike most other spur makers of the time. He learned the trade from a brother, who was also a cowboy. Oscar trailed herds until landing in Kansas City, Missouri, where he sought work as a blacksmith. He fashioned spurs and bits on his own time, selling them at horse sales. In 1914, he took on a partner creating C&G bits and spurs but sold the business to Charles Shipley in 1916, who also bought out George Bischoff. Crockett moved to Pawhuska, Oklahoma, soon after, and in January 1920 he and his uncle Brice started the Crockett Bit and Spur Co. By 1927, they were selling 5,000 pairs of spurs and as many bits. He married in 1932. As sole owner of the company, Oscar moved to Lenexa, Kansas, and then to Boulder, Colorado, in 1943, growing the factory to 125 employees and worldwide recognition. During his first five years there, Crockett produced 330,000 bits and spurs. Deciding to move back to Lenexa, he traveled between the business in Colorado and his farm in Kansas until suffering a fatal heart attack in June 1949. His widow kept the company going until selling it to James Renalde, owner of the Denver Metals Foundry, in 1951.

CROCKETT

Circa 1920s-1930s, bottle opener pattern spurs, inside marked "Crockett." Produced by the maker to loosely copy the iconic No. 3 pattern, 20-coin design, being produced at the time by Joe Bianchi. Bianchi's success with this design spurred copies by many of his contemporaries. Early Texas spurs with a bottle opener shank are more valued by collectors today than are those with a standard shank.

CROCKETT

Circa 1920s to 1930s, inside-marked Crockett spurs with swinging buttons and bottle opener shanks.

K&C

Rare, four-heart pattern spurs in mint condition. Original K&C marked spurs are always single-mounted.

PASCAL M. "P.M." KELLY (1886-1976)

Born in Childress, Texas, P.M. Kelly was taught to make spurs from a blacksmith from whom the Kellys rented a house. P.M. moved to Hansford County, Oklahoma, in 1907 and did blacksmith and mechanic work in addition to making spurs. In July 1910, he sold his shop and followed McChesney to Pauls Valley, where Kelly was soon head forger and the highest-paid employee. He and Tom Johnson Jr. were considered by McChesney to be his best craftsmen. In 1911, the two left McChesney and set up their own shop in Dalhart, Texas. Kelly adapted power equipment to speed up their work. Johnson preferred the hand-forged methods and left the partnership. He was married to Kelly's sister and died of kidney failure in 1915. Kelly completed Johnson's orders, not marking the work which was Johnson's style. Kelly expected his two brothers, Leo and Grady, to join the business and began marking his spurs "Kelly Bros." When Clyde Parker was hired until 1919, work was stamped "Kelly Bros. & Parker." The company made engines, tools and spurs and had an international trade. Their biggest customer for bits and spurs was the Puerto Rico Sugar Co., buying Kelly Bros. spurs and bits for their cowboys. Kelly moved his company to El Paso in 1924. During the Depression, he worked in Mexico manufacturing engines to power pumps for irrigation. His brothers ran the El Paso business. When Kelly returned in 1939, spur styles had changed drastically. Fewer cowboys and more rodeo performers were buying spurs. They preferred short shanks, wide heel bands and small rowels, causing Kelly to have to retool much of his equipment. He bought the company from his brothers and began naming the new styles for rodeo stars. After WWII, Bob Kelly joined his father, and the business was called Kelly and Sons. Business boomed for many years. In 1965, Kelly sold out to Jim Renalde of Denver, who had also purchased Crockett's spur business. Kelly moved to Oceanside, California, where his son, Jack, lived. By then, P.M. was nearly blind and died at age 90.

KELLY BROS.

KELLY BROS.
"Devil River Specials" with 1 1/4-inch heel bands and 3-inch, 6-point spur rowels. One-of-a-kind revolver mounted on the heel band was probably a later addition. The standard catalog pattern was mounted with a long triangle on the heel band. Fitted with Kelly's patented swinging buttons, this was one of the largest patterns he ever produced.

KELLY BROS.

KELLY BROS.

(Above) Rare, Kelly Bros.-marked Tom Watson pattern spurs. Fewer than five pairs of this pattern are known. Sometimes confused for the "Diamond Dick," which had a different shank and heel band mounting. The only catalog reference to the pattern is in the Kelly spur line sold through the Denver Dry Goods Co. in the latter 1920s through the first years of the 1930s. It is one of the most desirable of all his early patterns and the best pair of P.M. Kelly's work in the collection.

AL SMITH

CHAPTER III

EXAMPLES OF THE EARLY TEXAS STYLE MASTERS

1900-1950

EXAMPLES OF THE EARLY TEXAS STYLE MASTERS 1900-1950

Thomas A. Johnson Sr. & Tom Johnson Jr.
Charles P. Shipley
Joe Bianchi
Wallie Boone
Henry C. Zimmer
Jess Hodge
Al Smith

The James Wheat Collection contains fine type examples of a number of the Texas Style's greatest early makers. The most popular method of collecting spurs today is to acquire one or more examples of the important makers of the collector's regional style of interest.

The Early Masters of the Texas Style operated roughly between 1900 and 1950. Of this period's makers, Wheat acquired examples of work by Tom Johnson Sr., Tom Johnson Jr., C.P. Shipley, Joe Bianchi, Wallie Boone, August Buermann, Jess Hodge, Henry Zimmer and Al Smith. This was no small feat given the time period Wheat collected.

TOM JOHNSON JR.
Tom Johnson Jr. made these spurs patterned after the P.M. Kelly No. 56. Although never marked, Johnson's spurs are easily identified by their extra-measure of finish work, engraving style and close resemblance to Kelly's line.

TOM JOHNSON SR.

Rare pattern, Tom Johnson Sr. gal-leg spurs with alternating silver and copper dog bone mountings. This unmarked pattern is often confused with a similar design made during the period by G.A. Bischoff. The difference in value between the two makers is enormous.

TOM JOHNSON SR.

Gal-leg spurs mounted in copper and silver. Tom Johnson Sr. never marked his spurs or produced a catalog that is known. The majority of spurs attributed to him are in this pattern.

THOMAS A. JOHNSON SR. & TOM JOHNSON JR.

(1875-1915)

Tom Johnson Jr. was born in Arkansas, the son of his blacksmith father who came West from Tennessee. They worked together for a time in 1890 in Coleman, Texas, at Johnson & Son Blacksmiths. Both made spurs, but neither marked his work. Nothing more is known of Tom Johnson Sr. In 1905, Tom Jr. worked in Albany, Texas, making bits and spurs and sharing space with a saddle and harness worker. In 1910, he moved to Pauls Valley, Oklahoma, to work for J.R. McChesney. There he met P.M. Kelly. The two were considered McChesney's most talented craftsmen. Less than a year later in 1911, the two decided to leave Pauls Valley and set up their own shop in Dalhart, Texas, calling it Tom Johnson & Kelly, the Spur Makers. Johnson married Kelly's sister that same year. Tom was 36 and liked making spurs and bits the hand-forged way. P.M. was 25 and inventive. He figured out how to adapt mechanization to their products and increase their output. Tom left the partnership and established the Spur and Bit Factory in 1912, making only hand-forged products. Three years later, Tom died of kidney disease in 1915. According to Kelly, "Tom Johnson was one of the best and fastest hand-workmen ever." In Tom's hand-forged style, Kelly completed his former partner's orders, leaving them all unstamped. Lena Johnson died a few years later. The two had no children.

SHIPLEY

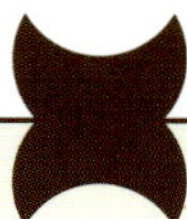

SHIPLEY

Two fine pairs of Shipley spurs that exhibit the unmistakable hand of Oscar Crockett who ran the Shipley shop in the late 19-teens.

S

CHARLES P. SHIPLEY (1864-1943)

Charles Shipley was a businessman who sold the bits and spurs of others in his catalogs and large Kansas City saddlery and mercantile company. Born in Wooster, Ohio, he opened at age 21 what became one of the earliest and most prominent businesses of its kind – the Charles P. Shipley Saddlery and Mercantile Co. of Kansas City. He married in his early 20s and built the three-story Shipley Building. He incorporated in 1910 and sold harnesses, boots, saddles, chaps, bridles and all types of leather goods. He featured the bits and spurs of such well-known craftsmen as J.R. McChesney and August Buermann, stamping the C.P. Shipley mark into each piece. Around 1915, he bought the Gainesville, Texas, company of George Bischoff and the Kansas City business of Oscar Crockett, setting up his own foundry for the production of bits and spurs. Following a tiff with McChesney, Shipley hired some of the spur maker's employees away. Bischoff worked for Shipley after selling him his business. Everything was to feature his C.P. Shipley mark. Crockett was appointed head of the bit and spur division of the vast mercantile business. But in 1920, Shipley closed it and sold his tools to Crockett, who established Crockett Bit and Spur Co. across the street from the Shipley Building. In 1940, Shipley suffered a stroke and died in 1943. His son and grandson continued the business until 1972.

SHIPLEY

Shipley's spurs were marked inside the heel band with "C.P. Shipley" on one spur and "Kan.City Mo." on the opposite. In a few instances, Shipley's spurs were not marked but can be identified by their design and workmanship.

SHIPLEY

(Above) Charles Shipley bought the Gainesville shop of G.A. Bischoff in 1915. This pattern is a variation of Bischoff Style R as shown in his 1911 catalog. Bischoff continued to work at the Shipley spur shop and incorporated many of his most popular designs into the Shipley line.

BIANCHI

Roughly 50 percent of Bianchi's spurs were marked. When marked, it is always inside the heel band. The most common mark is "Hand Forged" then "J. Bianchi Victoria, Tex." The rarest mark is "A.J. Fimbel." Fimbel was a saddlery in Victoria that sold Bianchi's products.

BIANCHI

Bianchi's No. 1 pattern was his most successful spur design, judging by the large number that have survived through the collector's market.

BIANCHI

(Right) Joe Bianchi No. 3 pattern, 20-coin spurs with trademark bottle opener shanks. Of Bianchi's four catalog patterns, this one on average is the most valuable.

JOE BIANCHI (1871-1949)

At age 14, Joe Bianchi left his home in Origgio, Italy, and moved with the family to Victoria, Texas, after his father convinced the family to join him. Joe and his brother, Paul, opened Bianchi Brothers Blacksmith Shop in 1891. Joe married in 1905 and separated from Paul so his brother could do general blacksmithing and Joe could specialize in bits and spurs. The spur business was located near his home in Victoria. It was not long before Bianchi spurs became sought after by ranchers and cowboys in the South and East and from as far away as Oklahoma, Oregon and California. He began marketing his work through pocket-sized catalogs and placing ads, becoming famous for his bottle opener shanks, Mexican silver coins, stationary buttons and unique rowel pin covers. His spurs are marked with either "Bianchi" on one spur and "Victoria Tex" on the other or "Hand Forged" inside the heel band of both spurs. Spur maker Raymond Bego was his protégé. Bianchi continued to make spurs, bits and branding irons until his death at age 78.

BIANCHI

"Make me a pair of those Victoria spurs." Bianchi's success with the bottle opener spur impacted spur fashion and design throughout the United States. More than 20 early makers offered designs featuring a bottle opener shank like Joe Bianchi's. The design remains popular today. (From left) Circa 1950s blacksmith-made; circa 1920s blacksmith-made; circa 1950s Crockett-Renalde; circa 1920s J.R. McChesney, No. 7 pattern. Oversize (center), circa 1920-30s Bianchi, No. 3, bottle opener in the heavy size. Marked "Hand Forged" inside the heel band.

WALLIE BOONE

WALLIE BOONE

Circa 1930s "Texas Centennial" pattern spurs with swinging buttons. This style is one of the most desirable of Wallie Boone's catalog patterns with collectors today. Listed as spur No. 59 in Boone's catalog No. 3, it sold for a handsome price of $4.50. Today, a fine pair can sell for 1,500 times as much.

B

W.R. "WALLIE" BOONE (1882-1958)

One of the Boone family of blacksmiths and spur makers, Wallie was born in Trent, Texas, in either 1880 or 1882. His father was blacksmith E.C. "Duff" Boone. Wallie learned spur making from his cousin, master craftsman and artist Bob Boone and worked in Duff's blacksmith shop starting in 1901. He opened his own bit and spur shop in Lubbock in 1924 with one employee. In the 1930s, he worked as a blacksmith in Whiteflat and in 1936 opened a factory in San Angelo. There with his brothers, Bigham, Jack and Bill, he made brads for Model T fenders, farm tools and bits and spurs. Boone used nickel and chrome to plate spurs, perhaps the first spur maker to do so. By 1940, he employed 15 men in the factory. Boone published a catalog and also sold through saddleries like Prosser Martin of San Angelo. He was said to be a true horseman who would exchange bits with a cowboy until he found the one that best suited the horse. In 1944, in a scuffle with a man he thought was trying to kill him, Wallie shot him and was sent to prison for five years. He was released in 1949, but was never the same again. A broken man, he moved to Andrews, Texas, to live with his sister who took care of him until he was put in a rest home and died in Rosco, Texas, in 1958.

WALLIE BOONE

(Left) Circa 1930s, with classic Boone engraving. (Right) Circa 1930s, Wallie Boone "Old Glory" pattern spurs.

HODGE

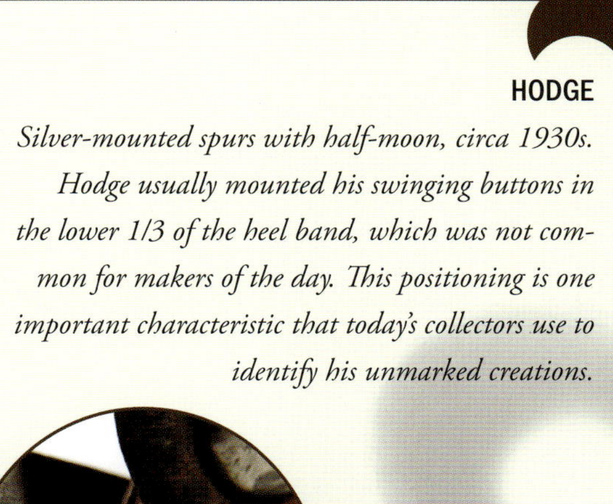

HODGE

Silver-mounted spurs with half-moon, circa 1930s. Hodge usually mounted his swinging buttons in the lower 1/3 of the heel band, which was not common for makers of the day. This positioning is one important characteristic that today's collectors use to identify his unmarked creations.

HODGE

Circa 1920s to 1930, with classic styling; fine multi-spoke, sawtooth rowels. This general pattern was Jess Hodge's most popular design. On some of his spurs, Hodge silver-mounted the iron where the top of the shank meets the top of the heel band. This mounting which forms a stylized "steer head," as it is known by collectors, is Hodge's only known mark.

H

JESS HODGE
(1870-1953)

This solitary man from Alabama was a mystery to those who lived in and around his shop in Fort McKavett, Texas. Hodge set up a business there in 1915 after having worked for the XIT Ranch in the Panhandle of Texas in 1912. In Fort McKavett, he had a 20- by 30-foot shop with a dirt floor that served as both his shop and his home. He slept there in a partitioned room that had a hard floor. His only companion was a dog. Hodge cooked on his forge and shared his meals with his dog, who ate out of the same pan after Jess had finished. There is speculation that his real name may have been J.H. Hodges and that he might have served prison time in Alabama, learning there to craft bits and spurs. He had a son in Oklahoma and two daughters in Alabama. Although his life was a secret, his artistry was unquestionable. In keeping with his desire to remain out of the mainstream, Hodge never marked his products. Their style and quality was easily recognizable, though, with their smooth finish and neat silver mounting. His spur shanks often had beveled edges, and he was known to use silver and copper from coins for mountings. His gal-leg spurs were famous, and he was known for creating bits that did not hurt a horse's mouth. He liked kids and often repaired their wagons and toys for some of their mothers' biscuits, bacon or fried potatoes. Hodge enjoyed playing dominoes and cards and drinking whiskey in the Trading Post Saloon with area ranchers. He had to quit all work except spur making due to arthritis. When he could no longer live alone, he moved into the Brady Nursing Home, supported by donations from Fort McKavett residents. When he died, no family members attended the funeral. A Methodist minister from nearby Mason, Texas, read "The Village Blacksmith" for the few friends in attendance.

ZIMMER

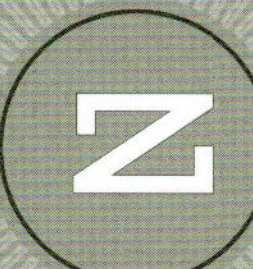

HENRY C. ZIMMER (1864-1934)

Henry Zimmer was born in Saint Francis County, Missouri, the youngest of nine children. His mom died when he was 2, and his farmer father died when he was 14. Henry worked in a brickyard then wanting a trade, he sought work with a blacksmith in Farmington, Missouri. At age 21, he headed to California but turned around and headed back, finding California to be unlike what he had expected it to be. Along the way, he stopped in Dallas and stayed, working as a journeyman blacksmith, having completed his apprenticeship in Missouri. A year later, Henry headed to Pecos, Texas, and set up shop in town quickly building a robust business. He married in 1891 and after years of blacksmithing and shoeing horses realized the automobile was changing the face of transportation. He opened a hardware store on his original blacksmith location and started selling cars – Chevys, Buicks and Fords. He and his son-in-law established Zimmer-Roddy Motor Co. and made a great deal of money. Henry served Pecos as county commissioner, Justice of the Peace and five years as mayor. He died of a heart attack the day after Christmas in 1934.

ZIMMER

Unmarked spurs attributed to Henry Zimmer. Unusual swinging buttons fashioned in the style of J.O. Bass. Zimmer produced spurs in Pecos very near where Wheat ranched and collected. Examples of this spur maker's work are rare.

ALFRED N. "AL" SMITH (1885-1954)

At age 9 Al Smith moved to Texas from his birthplace in Mound City, Kansas. His father set up a blacksmith business in Harrisburg near the busy Houston Ship Channel. Al helped his dad and made his home in Harrisburg for the rest of his life. (Houston annexed Harrisburg in the mid-1920s.) Al continued to live with his parents for several years and from 1922 to 1948 is listed as doing blacksmith work out of a shop on Cypress. He made distinct spurs with very narrow heel bands, light in weight, using flat Mexican silver coins and never marked his work. He is likely the first Texas spur maker to craft the bottle-opener shank design, although many attribute that to Bianchi in nearby Victoria, Texas. Likely, neither developed the design, and research indicates that the style was used in Mexico years earlier than when these two makers did their work. Also likely is that Smith had an influence on Bianchi, rather than the other way around. Smith was widely known for his light-weight, military type spurs, but he was never given the credit he deserved for his outstanding work. He collected the spurs of other makers during his life, but a theft of his shop in 1954 took most of his collection. The rest was sold by his wife after his death of a heart attack at age 69.

AL SMITH

Silver-mounted iron spurs with American coin-mounted buttons. This pair is exceptional with its extra-long shanks and finely made, 10-point rowels.

DEE BOONE

TEXAS STYLE SPUR MAKERS OF THE MIDDLE PERIOD

1940-1980

TEXAS STYLE SPUR MAKERS OF THE MIDDLE PERIOD

1940-1980

Adolph R. Bayers
Daniel "Dee" Boone
Leonard G. Grubb
White Deer
Huntsville Prison

During the 20th century, a signal change began to occur throughout the ranching community of the American West. The range cowboy of 1900 and the large outfits they were a part of began to give way to smaller, leaner operations utilizing more advanced ranching techniques that relied much less on the individual skill of the cowboy.

The rise of rodeo coincided with this decline of the open range, thus creating an entirely new venue and market for cowboys to showcase their skills. As the use and type of user began to change, so did styling in spurs. The basic design characteristics of a Texas Style spur remained the same but there were fashion-driven changes. Shanks became shorter and rowels, as a whole, smaller.

No middle period maker had a larger impact on the Texas Style spur than Adolph Bayers of Truscott, Texas. He was a living legend during his time, building a solid reputation for reliability, workmanship and fashion. Like the Early Masters who had come before him, he shaped and impacted the ever-evolving Texas Style of spur leaving his mark for generations.

The Middle Period also ushered in a whole new category of spur maker — the artist. Unlike the Early Masters and the Middle Period makers like Adolf Bayers, artists' spurs were, for the first time, beginning to be purchased primarily to collect and enjoy — not use. James Wheat's collection has several makers' spurs that fit this new classification. They typically are in mint, unused condition.

Of the Middle Period makers in the Wheat Collection, Adolf Bayers is the only one whose name was made by the cowboys who used them. Wheat collected spurs by the artists Dee Boone, L.G. Grubb, White Deer and the inmates of Huntsville Prison. The fine examples of L.G. Grubb spurs were likely ordered by Wheat as an addition to his collection. As might be expected, the Bayers spurs in the collection show some wear from use, while the artists' pairs are in unused condition. Adolf Bayers made contemporary designs that cowboys of his era wanted. The artists made designs, in some cases, reminiscent of the Early Masters or developed their own specific creations driven by their tastes alone. An artist could depend on a few patrons, like Wheat, who would buy multiple examples of his work with an eye to their uniqueness and artistry. The maker could only count on one purchase at a time, with future purchases driven only by the quality and durability of his product.

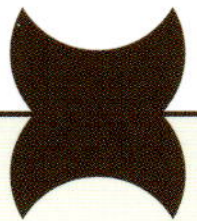

BAYERS

Bayers marked and recorded all but his earliest bits and spurs in this fashion. Today, thanks to the work of J. Martin Basinger and his books, collectors can reference Bayers' actual order ledger to verify the authenticity of surviving examples of Bayers' work.

ADOLPH R. BAYERS (1912-1978)

Born on a farm near Truscott, Texas, Bayers learned the art of spur making in his father's farm shop. He made his first pair of spurs in the 1930s, married in 1948 and continued living on the family farm. In the 1950s he began making spurs for polo players who liked his cowboy-looking designs over the more traditional English and military polo spurs. He began stamping his name and pattern number in 1952. With some of the equipment previously belonging to J.R. McChesney, Bayers developed artistic spurs noted for their exceptional silver work. Throughout his life Bayers kept meticulous records, and of all other Texas spur makers, only he and J.O. Bass preserved their original quotes and customer orders. Other than four years on the USS Kittyhawk in WWII, Bayers never left the family farm, dying there of cancer at age 65. Though his fame had spread far and wide, he always considered spur making his sideline job to farming. Much of Bayers' personal collection is in the Panhandle Plains Museum in Canyon, Texas. Bayers' original pattern books for his spurs and bits are in the collection of the National Ranching Heritage Center as a donation from J. Martin Basinger.

BAYERS

The most important Texas Style maker of the 1950s and 1960s was Adolf Bayers. His spurs were bought to use, with most surviving examples, like these, showing signs of wear.

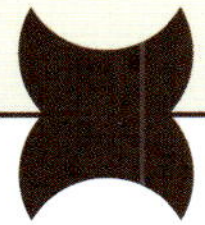

DEE BOONE

DEE BOONE

A fine collection of Dee Boone gal-leg spurs. He normally mounted his spurs in a combination of copper and silver and purchased stainless steel rowels rather than hand-making his own. Few pairs of Dee Boone's circa 1960s and 1970s production spurs show signs of use, leading one to the conclusion that most were originally bought by collectors.

DANIEL "DEE" BOONE (1898-1976)

Dee was born in Round Timber, Texas, one of several talented Boone spur makers. His father was Jerry Clayton Boone, and one of his cousins was spur maker Wallie Boone. Dee was living in Trent, Texas, when he made his first spurs. In 1915 living in Decatur, he left home to take part in the Wild West Show his brothers had formed. It moved from town to town until settling in Decatur and disbanding in 1922. Dee left the show after three years in 1918 deciding to set up a blacksmith shop. He and his dad moved to Pittsburgh, Oklahoma, where they did blacksmithing and made bits and spurs out of their house. Dee relocated to Henryetta, Oklahoma, in 1927 with his wife, Altha, a niece of entertainer Will Rogers. He set up a shop in town and did blacksmith work, also creating bits, spurs, knives and doing some gunmaking. After 45 years, Boone closed his shop but kept working out of his house. He died from a heart attack in Henryetta in 1976.

DEE BOONE

(Right and below) Dee Boone matching gal-leg bit and spur set.

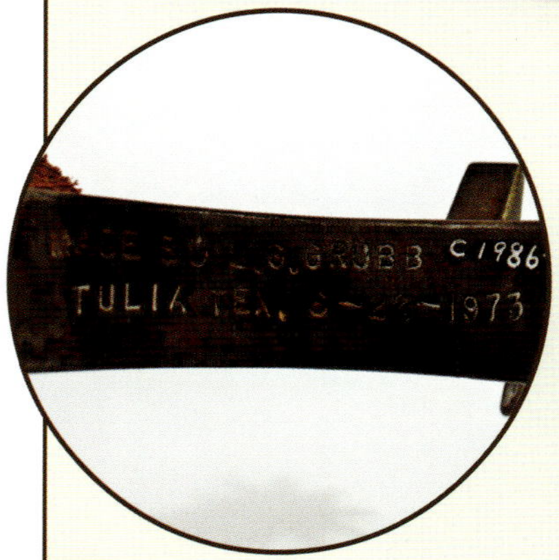

L.G. GRUBB

These brass inlaid spurs with swinging buttons, circa 1960s to 1970s, were made in Tulia, Texas. Grubb's spurs typically had swinging buttons and were well-marked with his name and the date they were produced.

G

LEONARD G. "L.G." GRUBB (1900-1989)

At age 6 months, L.G. Grubb came to Texas with his family from their home in Thomaston, Georgia. The family moved first to the Dallas area, eventually settling near Kress, Texas, where his dad farmed and did blacksmithing to maintain his equipment. L.G. enjoyed watching well-known spur maker J.O. Bass in Tulia as he crafted spurs and bits and decided to try making a pair himself. Gathering materials from what he could find, the young man created his first spurs in his father's shop. Enjoying the process, he made a second set, which he kept all his life. In 1934, he married and moved south of Tulia to farm. At age 60, L.G. quit farming and went back to spur making, working outdoors at his father's anvil and forge that had been brought to Texas from Georgia. Numerous orders for his good work kept L.G. busy for the rest of his life.

WHITE DEER

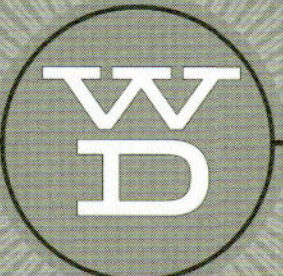

WHITE DEER

Biographical information on White Deer has not been found, creating speculation that the unknown spur maker may have stamped "White Deer" for the town of the same name or for the White Deer Rodeo. The White Deer spurs are mounted in brass and copper and were said to have been made in the Texas Panhandle in the 1950s.

WHITE DEER
The White Deer mark was inside the heel band as can be seen in this photo.

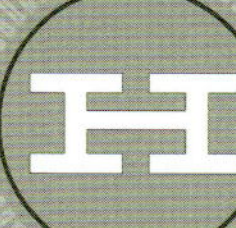

HUNTSVILLE PRISON

The Huntsville Penitentiary opened in 1849 and was the only Confederate prison standing at the end of the War Between the States. As a means of recreation for inmates starting in 1931, rodeos were staged with prisoners as performers. Several top riders of the early 1900s got their start in the Huntsville Prison Rodeos. Likewise, many fine pairs of spurs and bits were fashioned by inmates to be used in the rodeos and to raise funds for the prison. (It is known that spurs created years before the rodeos were a part of the Huntsville Prison, with one political gift set of spurs being made in 1910.) After 1938, stainless steel began being used. Typical prison spurs featured abalone mother of pearl and/or brass. They had straight shanks, rowels that were not sharp and often were decorated with U.S. dimes and quarters that were made in the 1920s and '30s. (Side note: Using U.S. coins in their designs was illegal for spur makers, but prisoners just bent the rule; outside spur makers had to use Mexican coins in their work.) Huntsville prisoners did not mark their work and few examples exist.

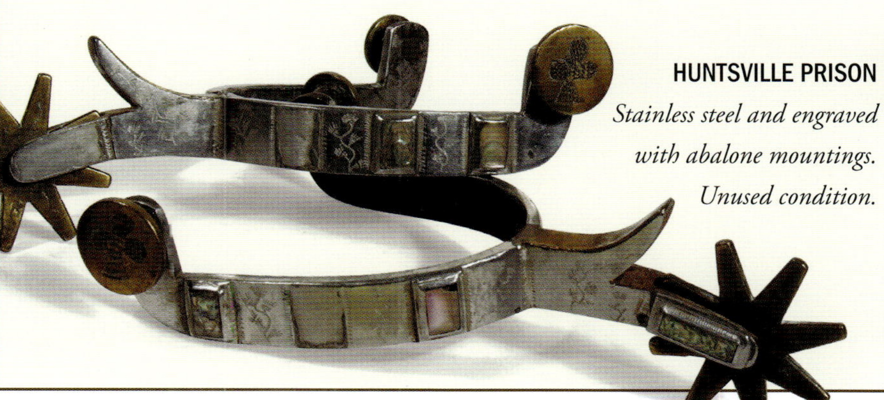

HUNTSVILLE PRISON
Stainless steel and engraved with abalone mountings. Unused condition.

HUNTSVILLE PRISON

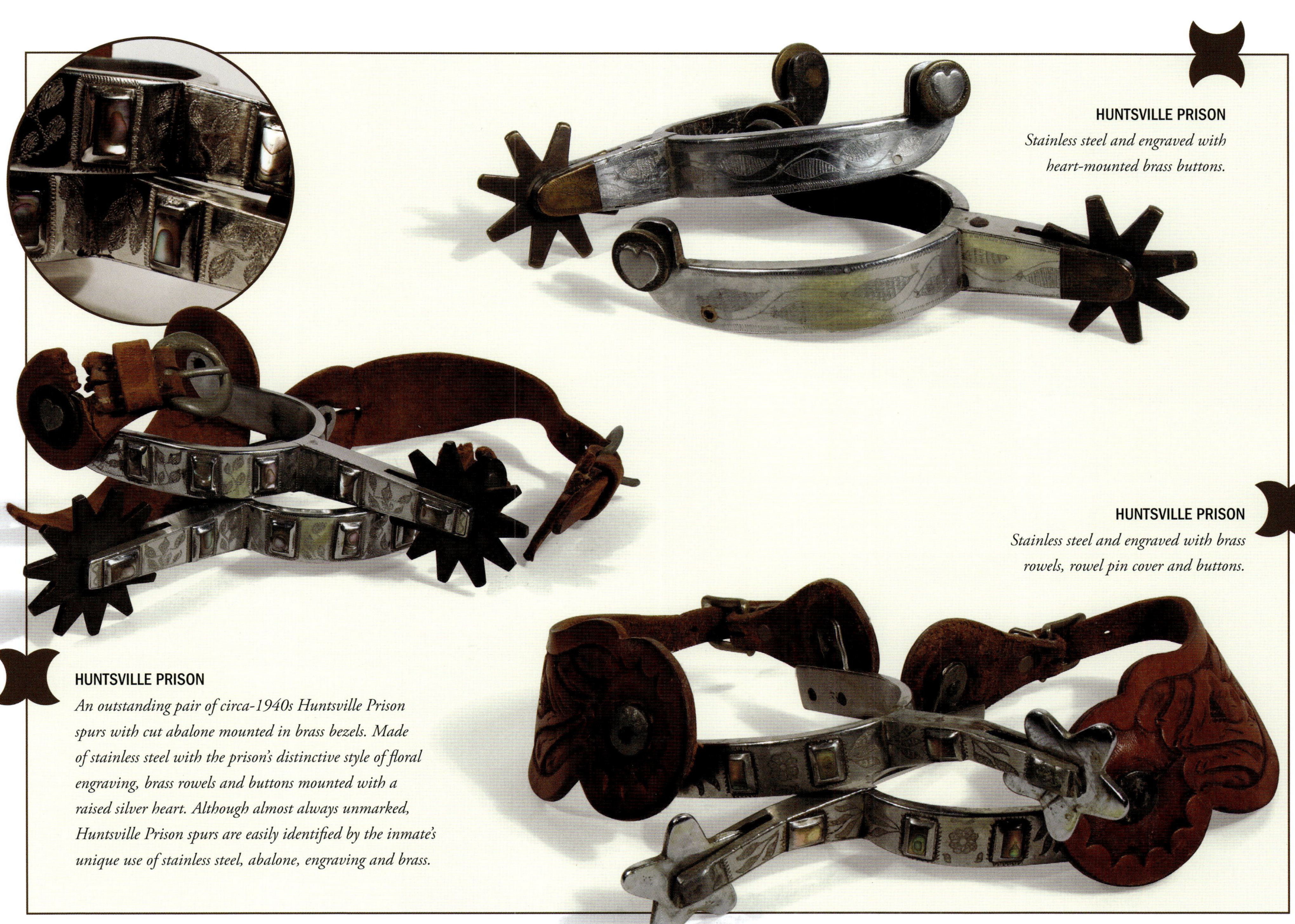

HUNTSVILLE PRISON

Stainless steel and engraved with heart-mounted brass buttons.

HUNTSVILLE PRISON

Stainless steel and engraved with brass rowels, rowel pin cover and buttons.

HUNTSVILLE PRISON

An outstanding pair of circa-1940s Huntsville Prison spurs with cut abalone mounted in brass bezels. Made of stainless steel with the prison's distinctive style of floral engraving, brass rowels and buttons mounted with a raised silver heart. Although almost always unmarked, Huntsville Prison spurs are easily identified by the inmate's unique use of stainless steel, abalone, engraving and brass.

KLAPPER

THE MODERN MAKERS

1970-PRESENT

THE
MODERN MAKERS
1970-PRESENT

The Wheat Collection contains three makers from the Modern Era that are still living today. Of the three, Billy Klapper and Ray Anderson are both makers and artists and Carl Hall an artist. Working cowboys and collectors alike have embraced the work of Billy Klapper and Ray Anderson. Both are considered today as two of the Texas Style's top living makers. With the growing popularity of spur collecting as a hobby since 1985, most top spur makers are also collected. There are still makers today, as in the Middle Period, who are almost exclusively collected and not used. Carl Hall is an example of this type of modern era artist.

Ray Anderson
Billy Klapper
Carl Hall

KLAPPER

Outstanding No. 327 gooseneck pattern, silver-mounted spurs. The split silver and copper heart mountings on the swinging buttons are in the style of J.O. Bass.

KLAPPER

An outstanding pair of Johnny Mullins-style Billy Klapper No. 317 spurs with "Cheyenne split" heel bands. This pattern was first made popular by Oscar Crockett in the 1920s.

BILLY KLAPPER (1937-PRESENT)

Born in Lazare, Texas, Billy Klapper grew up working as a cowboy on area ranches. In 1962, he was encouraged by the foreman of the Y Ranch to try bit and spur making. Klapper was most fond of Adolph Bayers' work, which was very popular with the ranchers and cowboys in his part of Texas. Bayers was not interested in sharing his methods nor designs and would not teach the young man spur making. Despite the lack of hospitality, Klapper was nevertheless influenced by Bayers' craftsmanship. After working as a cowboy by day and spur maker by night, Klapper moved to Childress, Texas, in 1968. He began marking his work with his name and style number, which he has continued throughout his career. When McChesney died and his equipment eventually sold to a junkyard by Nocona, Bayers bought it and used the tools until he died. Klapper bought the punch press, rolling machine and other equipment from Bayers' widow. In 1973, Klapper moved to Pampa, Texas, and continues his spur-making work, whittling away at the long waiting list of those wanting his beautiful bits and spurs.

KLAPPER

One of the most important of the modern Texas Style spur makers, Billy Klapper made these copper and silver-mounted gal-leg spurs.

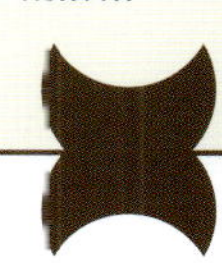

KLAPPER

The ultimate prize-the Klapper mark.

ANDERSON

RAY ANDERSON (1928-PRESENT)

Born in West Texas, Ray Anderson was a teen when he began helping his dad with forging and welding. Following service in the Army, Ray worked in the oil fields. He taught himself to make spurs in 1977, selling them at shows and auctions. He had collected the spurs of other makers but never talked with them about their techniques or getting pointers until he had made more than 200 pairs himself. Before moving to Abilene, Texas, Ray was making 30 pairs of Texas Style spurs each year while farming in Weatherford. His spurs are consecutively numbered; his bits are not numbered at all. Anderson is considered to be a very innovative designer. Contemporary spur maker Randy Butters says Anderson inspired his own work as a spurmaker.

ANDERSON

An early pair of Ray Anderson gal-leg spurs mounted with copper and brass with hearts on the heel band.

CARL HALL
(1937-PRESENT)

Carl Hall was born in Comanche, Texas, and learned the trade of blacksmithing from his father. When the young man saw the spurs of McChesney and Kelly Bros. he thought he would like to try making spurs, too. He favored the big Kelly Bros. designs. In 1964, Hall made his first pair of spurs and loved the process. He was especially pleased recently when he was able to buy back from the owner the third pair of spurs he'd ever made. Hall married in 1958 and runs the Southwest's largest mule and wagon sale, including bits and spurs, on the second weekend of every September and April. He continues to make spurs every day – 90 percent in the gal-leg design. He taught his son to make spurs and after that first pair the young man said, "I will never work that hard again," and he has farmed ever since. The Wheat Collection features Carl Hall gal-leg spurs, all in unused condition. According to dealers, most of his creations have gone into collections.

CARL HALL

Gal-leg spurs, all in unused condition. Most of Hall's creations are in collections instead of tack rooms.

CAÑON CITY PRISON

FOREIGN KINGS

FOREIGN KINGS

Guadalupe Garcia
Goldberg, Staunton
Miguel M. Morales
Cañon City Prison
Kurt Pfeiffer

The Wheat Collection contains fine type examples of a number of California and hybrid style Early Period spur makers. The addition of these makers' work was likely the result of Wheat's desire to have a collection that represented as many of the American makers as possible from throughout the West. He was limited in his attempt by what was available in the collecting network of his time. His geography, more than any other factor, determined what, for the most part, would be available to him.

G.S. GARCIA

Outstanding G.S. Garcia-marked, early California Bit.

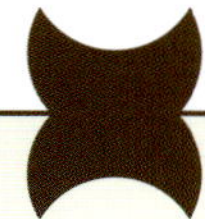

G.S. GARCIA

Garcia was the first American spur maker to command great value among collectors. Although his spurs were made in the California Style, it is not uncommon for collectors to include examples of his work in a Texas Style spur collection. Wheat would have considered himself fortunate to own these three nice pairs. When they were acquired, they would have been some of his most expensive additions.

GUADALUPE "G.S." GARCIA (1864-1933)

G.S. moved to California from his birthplace in Sonora, Mexico. He apprenticed in a saddlery in San Luis Obispo and learned to make saddles and spurs. In 1883, he opened his own shop in Santa Margarita, California, but due to the stiff competition, he moved to Elko, Nevada, in 1896 and started a business there. The high numbers of cowboys and ranchers in and around Elko provided a good clientele. Garcia's spurs were produced in the California Style, primarily, and were some of the most artistic of any spur maker of his time. Besides a local and regional appeal, Garcia spurs were popular in the Pacific Northwest, Argentina, Australia, Mexico and France. Garcia worked with dozens of apprentices, teaching them the art of spur making. Many became well known in their own right, among them Juan Estrada, A. Herrera, Raphael (Philo) Guttierez and Mike Morales. In 1912, he founded the Elko Rodeo, becoming a rodeo promoter. The first Elko Stampede drew 2,000 spectators. Garcia turned the spur making business over to his sons in 1932 due to failing health and retired to California, where he had first learned to make the beautiful, ornate spurs that made him famous. He died in North Hollywood in 1933 of a kidney ailment. The business was moved to Salinas, California, in 1936 and became the Garcia Saddlery.

GOLDBERG, STAUNTON

S

GOLDBERG, STAUNTON

Goldberg, Staunton was one of the largest retail businesses in Winnemucca, Nevada, going through several name changes over its 52-year life. Established in 1869 as M.B. Staunton Co., it was owned by Michael Buckley Staunton. In 1910 M.B. entered into a partnership with saddler Gus Goldberg to form Goldberg, Staunton Saddlery Co., specializing in bits, spurs, saddles, tack and Western and vaquero clothing. M.B.'s eldest son, David, who was born in 1880, worked in the mercantile and took over the business in 1900 at his dad's retirement. (M.B. died in 1905.) David was elected in 1904 to one term in the Nevada State Assembly. Around 1906, he expanded the bits and spurs division and employed talented silver worker Juan Estrada (1865-1942). David sold his interest to Goldberg in 1913 and moved to California. The business name changed in 1917 to Gus Goldberg Saddlery Co., but for a short period of time. Both Goldberg and Estrada moved to California in 1921 when the business was sold to James Otis. David Staunton died in San Francisco in 1939; Goldberg and Estrada died in Sacramento in 1942.

GOLDBERG, STAUNTON

This extremely rare pair of marked Goldberg, Staunton, Winnemucca, Nevada, spurs is in a popular pattern that was produced by a number of the era's California Sytle makers. A marked pair by this maker is extremely difficult to obtain.

MORALES

MORALES

Mike Morales enjoyed a storied career that began in the shop of G.S. Garcia in Elko, Nevada, in 1902. This scalloped and inlaid pair is marked with Mike Morales' squashed M and were probably made in the 1920s.

MIGUEL M. "MIKE" MORALES (1888-1934)

Perhaps the most prolific of the important California Style makers was Mike Morales. Born in Mexico, he moved to Elko, Nevada, in 1902 to apprentice with G.S. Garcia until 1906. He moved at age 22 to Pendleton, Oregon, and opened a bit and spur department for Hamley & Co. Accepting as many as seven Mexican apprentices, fights and absences were common. Morales had a short temper, as did some of the students. At one point, he sold his tools to an apprentice who a year later sold them to another apprentice, Irl Terry. Irl closed the business and moved to California and made bits and spurs for Hamleys in Sacramento. Morales, in 1915, moved to Portland, Oregon, and developed a unique and distinctive line of California Style bits and spurs before opening his own shop there in 1916 – M. Morales and Co. Morales marked much of his work with a squashed M. In the meantime, he married and divorced and moved to Los Angeles in 1927. He established a business on Sunset Boulevard and hired a craftsman to work with the tools, dies and metalwork. Morales handled the designing and engraving. He died in 1934 of a streptococcal infection. He was considered one of the most prominent and recognizable of the Elko school spur makers.

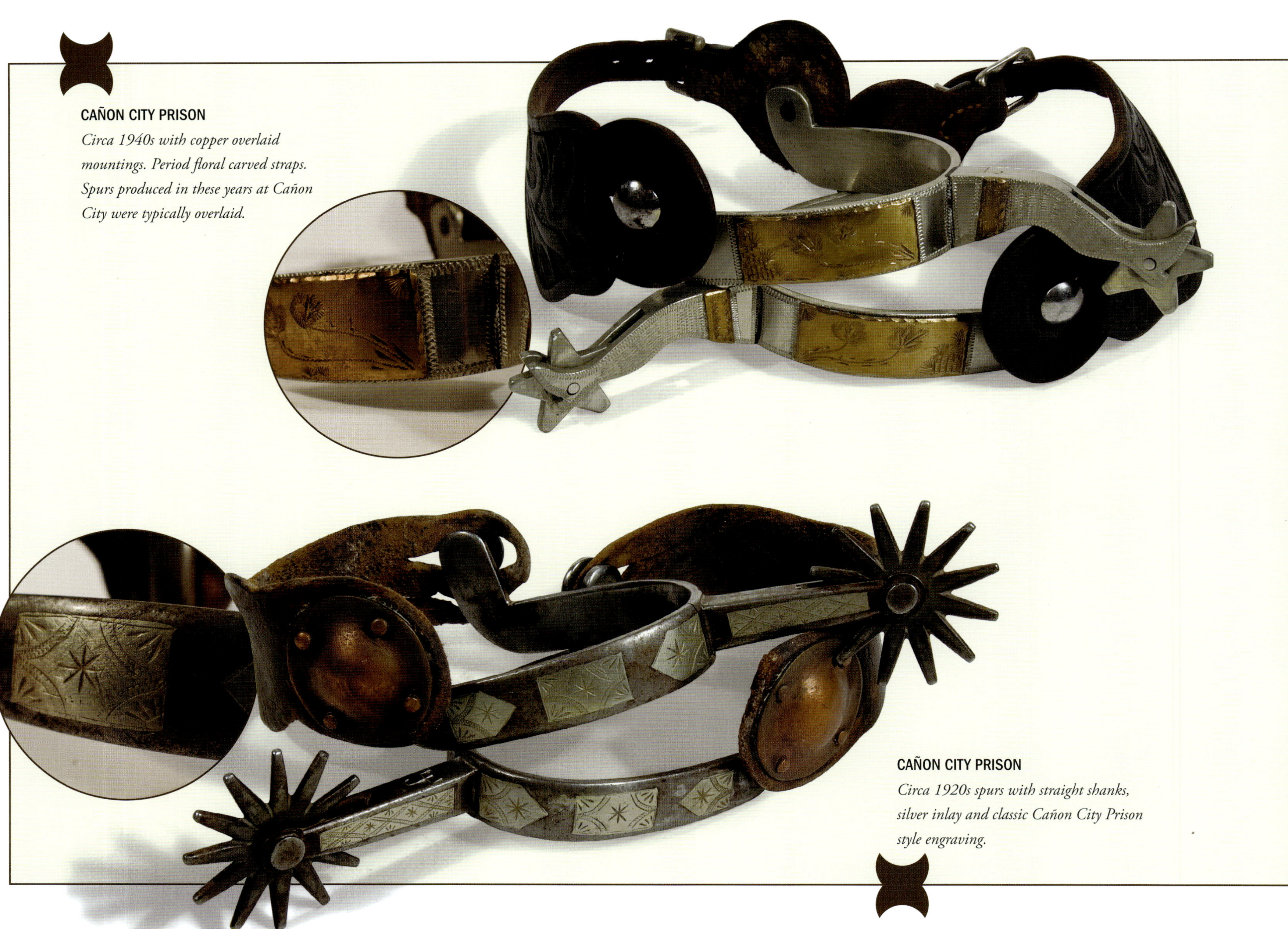

CAÑON CITY PRISON

Circa 1940s with copper overlaid mountings. Period floral carved straps. Spurs produced in these years at Cañon City were typically overlaid.

CAÑON CITY PRISON

Circa 1920s spurs with straight shanks, silver inlay and classic Cañon City Prison style engraving.

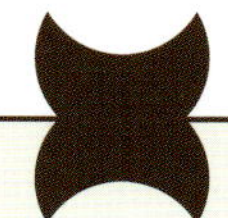

CAÑON CITY PRISON

Silver inlaid bit with military-inspired S shank cheek pieces.

CAÑON CITY PRISON

Many bits and spurs were made at the Colorado State Penitentiary. According to Maul & Ferguson's Cañon City Spur Book, the best spur makers of the time were contemporaries whose sentences overlapped during the 19-teens and '20s. They were incarcerated together in Old Cellhouse #4 during the tenure of reform warden T.J. Tynan who offered them access to tools and torches in addition to encouragement and support. Distinctive characteristics include the prisoners' numbers stamped on some spurs, double-mounted nickel silver inlays with identical patterns on both sides and geometric or vine and leaf engraved patterns. Some of the prisoners, when released, continued their spur making in the shops of others or in their own trades.

CAÑON CITY PRISON

Circa 1920s silver inlaid "Slipper" pattern spurs with bird head shanks. This popular pattern was produced a number of times over the years and was sometimes made with cutout heel bands. The few Cañon City Prison spurs and bits that were marked were only done so with the inmate's prisoner number.

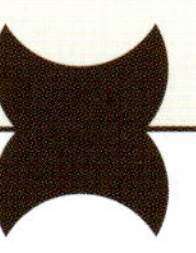

PFEIFFER

KURT PFEIFFER
(1893-1983)

Kurt Pfeiffer was born in Colorado and blacksmithed there in the towns of Silt and Rifle. Although beautifully crafted, his bits and spurs were few in number because Pfeiffer's wife had a strong distaste for his time being wasted on work that paid so poorly. She preferred he spend his endeavors in traditional blacksmithing. The state of Colorado had few bit and spur makers at this time, and with the quality of his work being what it was, no doubt his talent would have been in demand. Unfortunately, most spurs and bits sold for only a few dollars. Pfeiffer died in Santa Rosa, California, but research reveals very little else about this skillful spur maker.

PFEIFFER
Rare Kurt Pfeiffer marked bit. The state of Colorado had few bit and spur makers at the time Pfeiffer crafted his works.

THOUGHTS ON VALUE AND CONDITION

The three most important factors that influence a spur's value are: The maker, the pattern and the condition.

THE MAKER is the most important element in determining the value of a pair of collector spurs. There were more than 100 well-known spur makers who produced antique American Western spurs. Some were prolific, while others created few pairs in comparison. Over time, the value of the individual maker is determined by the market. As a rule, the greater the artistry and craftsmanship of the maker, the more desirable are his creations today. The maker's exposure in books and articles also greatly impacts value, as does rarity.

THE PATTERN of a pair of early collector spurs is an important driver of value. Most all the early makers made a wide range of products designed to meet every market and budget. Manufacturing elaborate designs required more time and in some instances valuable materials. As a result, patterns that cost more to produce are much more valuable and rare today. There were fewer buyers in the 1920s who could afford a pair that cost $15 when the average pair was $5. There were also fewer people who would have wanted them, too, just like some today who can afford a Rolls Royce but choose a Chevrolet instead. As a result, the lower-priced spurs, which were plainer, sold in greater numbers. Fewer elaborate spurs were produced, making them rare and highly valuable to collectors today. As an example, P.M. Kelly's extremely rare (1 pair known) Round-Up Special pattern, which cost $15 in his 1920 catalog, command up to $75,000 today. The average pair of early Kellys today sells for around $1,200. One other important factor is the relative rarity of the pattern within a maker's line regardless of how fancy the pattern may appear. Some collectors attempt to own all the different styles made by a specific maker, in which case rarity of the pattern alone could drive up its value.

THE CONDITION of early spurs greatly affects the value. Spurs were originally created to be worn, so few antique spurs are found in pristine condition. Spurs reflecting little to no wear will command a much higher price than spurs that show use — more of the spur's original artistry can be seen when the spurs are in better condition. A spur's condition can be assessed from totally reworked to pristine. A pair of spurs in pristine condition can be worth as much as 20 times more than the same pattern that has been completely reworked. If you have what might be a valuable early pair of spurs, never clean or attempt to restore the pair without first being advised by a knowledgeable collector, dealer or restorer. The Wheat Collection contains seven examples of P.M. Kelly's legendary #18 pattern spur, all in varying conditions. The condition scale can be applied to these spurs for value comparison.

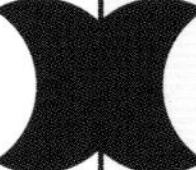

USED HARD

(Opposite page, top) A grouping of fine and rare early Texas spurs in as-found condition. The first four pairs are Kelly Bros., circa 1915 to 1935; the pair on the right are by Oscar Crockett, circa 19-teens. If these spurs were in excellent, all-original condition, they would be worth as much as five times more.

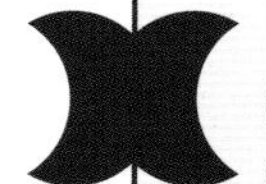

KELLY BROS. #18
ON A SCALE OF 1 TO 10

(Opposite page, bottom, from left)

1 *Re-rowelled, completely re-silver-mounted, cleaned. The iron is the only thing that is original Kelly Bros.* ***1***

2 *Repaired, poorly re-rowelled, slightly cleaned.* ***6.5***

3 *Missing one rowel pin cover, original patina, rowels blunted.* ***7***

4 *Missing rowel pin covers, repaired, patina cleaned, buffed off.* ***7***

5 *Missing rowel pin cover, button cover, rowels repinned in period use; could be easily restored, undisturbed patina.* ***8***

6 *All original, undisturbed patina.* ***8.5***

7 *Excellent, all original; fine, undisturbed patina, light wear.* ***9***

1
2
3
4
5
6
7

ACKNOWLEDGEMENTS

WITH MUCH APPRECIATION AND THANKS
TO THE FOLLOWING FOR HELPING WITH
THE DEVELOPMENT AND PUBLICATION OF THIS BOOK.

- The Diamond M Foundation, San Angelo, Texas, for funding the production of this book and making its publication possible.

- Bruce Bartlett, San Antonio, Texas, the fastest spur identifier in the West! The expertise of this man in his business of knowing and appraising Old West antiques is amazing to observe. Thanks go to Bruce for working on the identification and selection of spurs to be featured in the book, photo ordering, chapter writing and photo IDs in very short time. Thanks, too, to Julianne and Winn Bartlett for sharing him during back-to-back travels to Cody, Wyoming, for a big spur show, then on family vacations where he had no computer access for writing or e-mail. Bruce's text was done by hand and completed on deadline. All of the people involved with this project were in awe of his knowledge and very appreciative of the cooperation Bruce displayed in the preparation of this book.

- Joe Flores of Stratford, Texas, for his knowledge of spurs and spur people and his support and humor throughout the preparation of this book and the accompanying exhibit. He is a best friend and important resource for the National Ranching Heritage Center.

- Brandi Price and Mark Hartsfield, Hartsfield Design, for their excellent photography and graphic design of this book and the companion exhibit. Their expertise speaks for itself, but their professionalism and congeniality while working on a very difficult task are appreciated very much by the NRHC staff.

- Frankie Gongorda and Richelle Detrixhe of Hartsfield Design for their help in the production of this book, Carla Tedeschi for her ideas and assistance with creative photography and Duward Campbell for his expertise in photographing spurs.

- Robert W. Tidwell, NRHC curator, and R.G. de Stolfe, staff assistant, for the on-site help they provided to Bruce and the Hartsfield group.

- Ned and Jody Martin and Kurt House, authors of the incredible "Bit and Spur Makers in the Texas Tradition" and "Bit and Spur Makers in the Vaquero Tradition." Their books were the final word on the spur makers' biographies, as we attempted to verify facts and findings.

- Ray Westbrook of Lubbock, Texas, for background information on James Wheat. Not exactly easy to find, some of James Wheat's story was obtained with the help of this Lubbock Avalanche Journal reporter as he prepared for a story about the NRHC's spurs for the Lubbock Magazine.

- Ann Kathryn Orsinger who reflected Bruce Bartlett's expertise on spur collecting in the Collector's Guide of the April 2008 Cowboys and Indians magazine.

- Charles Kuralt's "On the Road with Charles Kuralt" book, the chapter titled "Loving County (Texas)," page 175. The book was published by G.P. Putnam's Sons, New York, NY © 1985 by CBS Inc.

- Other material found on the Internet, that wonderful source for information about everything.

- Jim Pfluger, executive director, for writing the Preface and Suggested Readings and whose knowledge of books and authors was a great help.

- Spence Miller, exhibit designer, for taking the Wheat book into an excellent exhibition for the National Ranching Heritage Center.

- And to Preston Lewis, an exceptional thinker and writer, whose good judgment is always valued.

Marsha Pfluger
Project Director
NATIONAL RANCHING HERITAGE CENTER

REFERENCES AND SUGGESTED READING

House, Kurt. *Hand Forged For Texas Cowboys.* San Antonio, Texas: Three Rivers Publishing Co., 2001.

Jacobs, Lee C. *J.R. McChesney – A Lifetime, A Legacy.* Colorado Springs, Colorado: Minuteman Press, 2006.

Manns, William and Elizabeth Clair Flood. *Cowboys and The Trappings of the Old West.* Santa Fe, New Mexico: Zon International Publishing Co., 1997.

Martin, Ned and Jody and Kurt House. *Bit and Spur Makers in the Texas Tradition.* Nicasio, California: Hawk Hill Press, 2000.

Martin, Ned and Jody. *Bits and Spurs: Motifs, Techniques and Modern Makers.* Nicasio, California: Hawk Hill Press, 2003.

Martin, Ned and Jody. *Bit and Spur Makers in the Vaquero Tradition.* Nicasio, California: Hawk Hill Press, 1997.

Maul, Patrice and Jack Ferguson. *Cañon City Spur: Colorado Prison Spurs and the Men Who Made Them.* Colorado Springs, Colorado: Old West Trading Co., 1995.

Miller, Ben and J. Martin Basinger. *Artistry in Silver and Steel: The Adolph Bayers Legends Vol I.* Slaton, Texas: OTO Co., 1996.

Miller, Ben and J. Martin Basinger. *Artistry in Silver and Steel: The Adolph Bayers Legends Vol II.* Slaton, Texas: OTO Co., 1997.

Pattie, Jane. *Cowboy Spurs and Their Makers.* College Station, Texas: Texas A&M University Press, 1991.

Pattie, Jane and Tom Kelly. *Cowboy Spur Maker – The Story of Ed Blanchard.* College Station, Texas: Texas A&M University Press, 2002.

Witherell, Brad and Brian. *Old West Art and Antiques.* Atglen, Pennsylvania: Schiffer Publishing, 1999.

INDEX OF PHOTOS